The INSIDE GUIDE

EARTH'S EXTREME PLACES

Polar Regions

By Judy Thorpe

Cavendish Square

Published in 2026 by Cavendish Square Publishing, LLC
2544 Clinton Street Buffalo, NY 14224

First Edition

Website: cavendishsq.com

Library of Congress Cataloging-in-Publication Data

Names: Thorpe, Judy, author.
Title: Polar regions / Judy Thorpe.
Description: Buffalo, NY : Cavendish Square Publishing, 2025. | Series: The inside guide: Earth's extreme places | Includes bibliographical references and index.
Identifiers: LCCN 2024059507 | ISBN 9781502674470 (library binding) | ISBN 9781502674463 (paperback) | ISBN 9781502674487 (ebook)
Subjects: LCSH: Polar regions–Juvenile literature.
Classification: LCC G614 .T467 2025 | DDC 910.911–dc23/eng/20250131
LC record available at https://lccn.loc.gov/2024059507

Editor: Jennifer Lombardo
Copyeditor: Nicole Horning
Designer: Deanna Lepovich

The photographs in this book are used by permission and through the courtesy of: Cover murathakanart/Shutterstock.com; p. 4 Jane Rix/Shutterstock.com; p. 6 PorcupenWorks/Shutterstock.com; p. 7 OSweetNature/Shutterstock.com; p. 8 StefanRenner/Shutterstock.com; p. 9 Soleil Nordic/Shutterstock.com; p. 10 KARYI YEAP/Shutterstock.com; p. 12 Maximillian cabinet/Shutterstock.com; p. 13 Christopher Teixeira/Shutterstock.com; p. 14 Risto Raunio/Shutterstock.com; p. 16 vladsilver/Shutterstock.com; p. 18 slowmotiongli/Shutterstock.com; p. 19 Thomas J Mitchell/Shutterstock.com; p. 21 David Osborn/Shutterstock.com; p. 22 maks_ph/Shutterstock.com; p. 24 Himmel-S/Shutterstock.com; p. 25 AB Forces News Collection/Alamy Stock Photo; p. 27 Colin Harnish/Shutterstock.com; p. 28 (top) Aris Group/Shutterstock.com; p. 28 (bottom) rbrown10/Shutterstock.com; p. 29 (top) axily/Shutterstock.com; p. 29 (bottom) muratart/Shutterstock.com.

CPSIA compliance information: Batch #CSCSQ26: For further information contact Cavendish Square Publishing LLC at 1-877-980-4450.

Printed in the United States of America

CONTENTS

The North Pole is surrounded by floating ice sheets that are thick enough in most places for animals and people to stand on.

Chapter One

EXTREME COLD

The polar regions get their name because they **describe** the areas around Earth's North and South Poles. The northern polar region is at the top of the world, in the middle of the Arctic Ocean, and is sometimes called the Arctic for this reason. It is made up of huge pieces of floating ice. The southern polar region is at the bottom of the world, on the continent of Antarctica.

The North Pole

The ice sheets at the North Pole are constantly moving, so no people live exactly at the North Pole. However, parts of Canada and Greenland are generally considered to be included in the northern polar region. People have lived in those parts of the world for thousands of years.

> **Fast Fact**
>
> People generally think of north as "up" or "top," and south as "down" or "bottom." Nearly all maps are **oriented** this way. However, this is a social construct, or human-made idea.

While most parts of the world have four seasons, the poles have only two: winter and summer. The average winter temperature

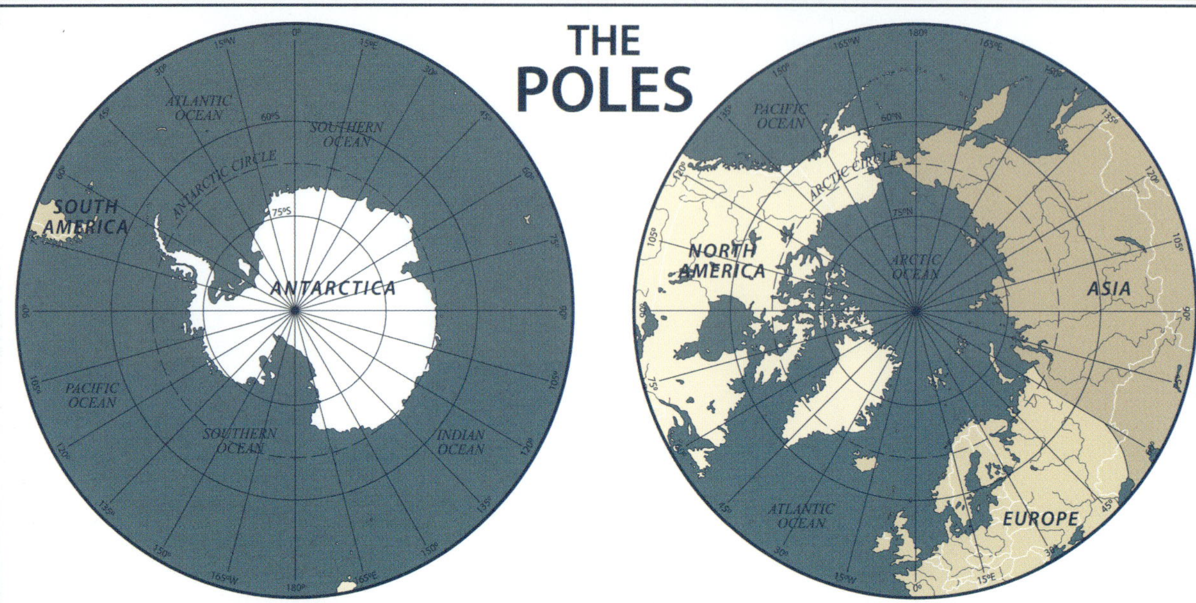

South Pole

North Pole

This picture shows the geographic locations of the poles.

Fast Fact

The North and South Poles are deserts! A desert is any area that gets less than 10 inches (25 cm) of rain per year. The poles have a lot of snow and ice, and even some liquid water, but very little rainfall.

at the North Pole is between –22 and –40 degrees Fahrenheit (–30 and –40 degrees Celsius). Summers are also very cold, with temperatures rarely rising above 32°F (0°C).

The South Pole

When people think of the southern parts of the world, they often think of warmth. However, the southernmost part of the world is even colder than the northernmost part! The average winter temperature at the South Pole is –76°F (–60°C), and the average summer temperature is –18°F (–28.2°C).

There are several reasons why the South Pole is colder than the North Pole. One is the continent of Antarctica is covered with thick ice

GEOGRAPHIC AND MAGNETIC

There are actually two North and two South Poles: a geographic pole and a magnetic pole. The geographic poles are determined by drawing an imaginary vertical line down the center of the planet, **perpendicular** to the equator, which divides Earth in half horizontally.

Earth is surrounded by a magnetic field that is caused by the liquid iron in Earth's outer core. If the geographic poles can be described as opposite ends of an imaginary line, or axis, the magnetic poles can be described as opposite ends of an imaginary bar magnet. The magnetic poles drift a little bit over time, which is part of the reason why they don't line up exactly with the geographic poles.

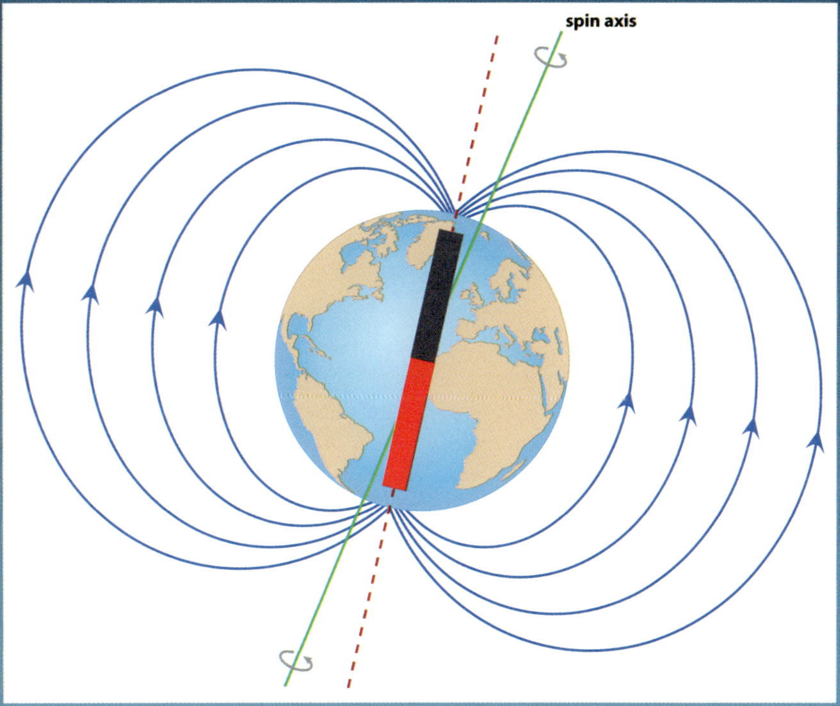

This picture shows Earth's magnetic field coming out of the magnetic poles, which the red dotted line passes through.The green line passes through the geographic poles.

Fast Fact

At the South Pole, it is summer from October to March and winter from March to October. At the North Pole, summer is even shorter: It runs from June to September.

that raises it above sea level. There are even mountains on some parts of the continent. The higher an area is above sea level, the colder the air gets.

Antarctica is also an area of land surrounded by water. When the wind blows, there is no land around it to slow it down so it has time to warm up. This means there are often very strong, cold winds blowing around the South Pole. In contrast, the North Pole is an area of water surrounded by land. The land around it slows the winds down

Shown here are some of Antarctica's mountains.

so they aren't as strong and cold.

A major reason for the **intense** cold at both poles is that the North and South Poles are dark for half the year. They are at such high latitudes that in the winter, the sun barely rises above the horizon. Without the sun's warmth, the poles stay very cold.

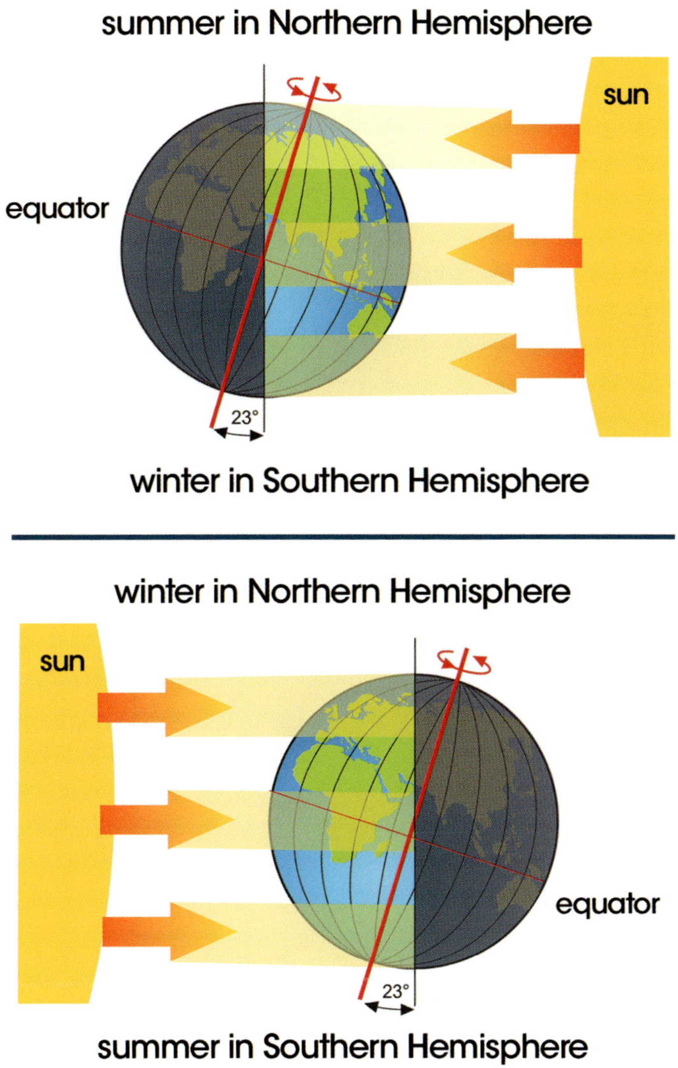

Yet another reason for the poles' extreme cold is that they do not get much direct sunlight. This picture shows the angle at which sunlight strikes the poles. It also shows how Earth's tilt toward or away from the sun creates different seasons in different parts of the world.

Narwhals are sometimes called "unicorns of the sea." A narwhal's horn is actually one long tusk, or tooth, similar to an elephant's tusks. Narwhals are one of only three species, or types, of whale that live in the waters of the Arctic Circle. The other two are belugas and bowheads.

NORTH POLE WILDLIFE

Chapter Two

The North Pole is water covered in pack ice. Pack ice describes thick, floating slabs of ice made by frozen seawater. Pack ice is not attached to land, but in some places it can cover the water enough to walk on it as if it were land. The living things that call the North Pole home have adapted to live on pack ice.

Fast Fact

There are many different kinds of pack ice, each with their own name. For example, a floe is a mostly flat piece of ice that measures 65.5 feet (20 m) or more across. Close pack ice is made up of multiple floes touching each other in some places, with water visible between them. Consolidated pack ice is made up of floes that are frozen together so no water is showing.

Polar Bears

The **iconic** polar bear is found only at the North Pole, not the South Pole. The polar bear is both the largest bear and the largest land predator in the world. Its main source of food is seals, especially bearded seals and ringed seals. Seals have high amounts of fat in their body, and polar bears need a lot of fat to fuel them so they can keep themselves warm and have enough energy to hunt.

This picture of North Pole pack ice was taken in 2016. This type is called "hummocked ice," which means the pieces pile up in a way that forms small hills.

Fast Fact

Polar bears have so much vitamin A in their liver that eating it could kill a human! If a person is lucky enough to survive, they may experience hair loss, dry and peeling skin, headaches, and weakened bones.

Polar bears have several **adaptations** that help them **thrive** in the Arctic. For example, although they look white from a distance, their fur is actually clear and hollow, while their

This polar bear's black skin can be seen in places where its fur is wet or sparse.

skin is black. The hollow hairs and dark skin trap more of the sun's light and heat to keep the bear warm. Another adaptation is their swimming ability. Polar bears need to be great swimmers so they can catch the seals they rely on for food.

Walruses

The walrus is another animal found only in the Arctic. Walruses have no fur to keep them warm, although they do have a layer of short hair. The main thing that keeps them warm is a thick layer of fat called blubber.

Walruses spend most of their time in shallow water, diving for food such as clams, snails, shrimp, and sea cucumbers. However, they need

MELTING ICE

Climate change is a huge threat to the North Pole. As the planet gets warmer, the polar ice melts more in the summer than it ever has in the past. Winters are also warmer on average, which means less ice reforms in the Arctic. Without pack ice, animals such as polar bears and walruses have to swim longer distances to find places to rest.

Melting Arctic ice also poses a threat to the rest of the planet. When there is a lot of ice, the difference in air pressure keeps the cold air in the north and the warm air in the south. With less ice, this pressure is weaker, allowing streams of icy air to travel south. This is why places as far south as Florida have sometimes experienced much colder winter temperatures than normal.

Sometimes several walruses will rest on one floe together.

to rest on solid ground between dives. They prefer to do this on ice floes. Walruses have flippers instead of arms, which are great for swimming but not strong enough to pull them onto the floe. Instead, they use their large tusks to pull themselves out of the water.

Arctic Foxes

Arctic foxes prefer to stick to the more southern **tundra** region of the Arctic, but they have been known to travel nearly to the North Pole in search of food. They are kept warm by their very thick coat of fur. To help them blend in with their surroundings, their fur changes color. It is gray or brown in the summer, when they stay mainly in the forests of the tundra. In the winter, when they are often spotted on ice floes, their fur is pure white.

Fast Fact

The pads of an Arctic fox's feet are not covered in fur. To keep them from freezing in the snow and ice, the foxes have a specially adapted **circulatory system** in their feet.

If the polar bear is the North Pole's most iconic animal, the emperor penguin is the South Pole's. It can grow to be nearly 4 feet (1.2 m) tall, making it the world's largest penguin species.

SOUTH POLE WILDLIFE

Chapter Three

Although the South Pole is located on solid ground rather than pack ice, that land is itself surrounded by water. For this reason, like North Pole animals, many of the animals that live at the South Pole spend much of their time in the water. Some, such as penguins and seals, spend time both on land and in the ocean. Others, such as whales, are found only in the ocean.

Penguins

Emperor penguins are the most popular Antarctic penguin, but they are the least common on the continent. Other species include gentoo, chinstrap, macaroni, king, royal, and rockhopper penguins. All penguins are flightless and are excellent swimmers. They have oily feathers that keep them warm and dry as they swim through the freezing waters, hunting fish and krill.

Fast Fact

There are 18 species of penguins in the world. Eight of them live in Antarctica. Two of those eight species live there permanently, while the others visit to breed.

Shown here are Adélie penguins jumping off the ice and into the water.

Fast Fact

Penguins come from dinosaurs! Avian, or bird, dinosaurs were the only dinosaurs that survived after a meteorite hit Earth millions of years ago. All birds alive today have **evolved** from those survivors.

Whales and Seals

Cold water can hold more oxygen than warm water, and the South Pole, like the North Pole, experiences 24 hours of sunlight for several months of the year. This combination allows small plants and animals, such as plankton and krill, to thrive. The huge number of South Pole microscopic creatures attracts larger animals that feed on them. This makes the waters around Antarctica the largest feeding ground in the world for marine animals, which live in the ocean.

There are 10 different species of whales found near the South Pole. Six are baleen whales. They are named after the flexible hairs that filter the whales' tiny prey from the water. These six species are the fin, southern right, sei, humpback, minke, and blue whales. The blue whale is the largest known animal to have ever lived on Earth.

The other four are toothed whales, which eat larger animals such as fish, squid, and seals. These four species are the orca (also known as the "killer whale"), sperm whale, Arnoux's beaked whale, and the southern bottlenose whale.

Humpback whales are frequently seen in the waters around the South Pole.

BERGMANN'S RULE

Why are so many polar animals so big? The answer is known as Bergmann's Rule, which states that some species of birds and mammals get bigger as temperatures get colder. Many scientists believe larger animals, which have a lower ratio of surface area to volume, are able to save their body heat better than smaller animals.

Not all scientists agree with Bergmann's Rule because it has so many exceptions. For example, although polar bears are bigger than other kinds of bears, Arctic foxes are one of the smallest species of foxes in the world. Some scientists believe there are reasons other than heat conservation that explain Bergmann's Rule.

Whales are known for their intelligence. Orcas, in particular, have displayed very intelligent hunting behavior. For example, they have been observed, or seen, creating large waves to wash seals off ice floes and into the water. Orcas are typically not found at the North Pole because the pack ice makes it difficult for them to come to the surface to breathe. However, as global warming melts the ice sheets, orcas are moving further north.

Fast Fact

Very few plants grow at either pole. On the North Pole pack ice, there are no plants at all. In the Arctic tundra around the pole, grasses and small bushes called shrubs grow in the summer. Most of Antarctica is covered in ice all year, but in some places, mosses and lichens are common.

This is causing problems as they attack animals that are not used to having them as predators—including other whales.

There are six species of seals that live in Antarctica: Antarctic fur seals, leopard seals, Ross seals, southern elephant seals, crabeater seals, and Weddell seals. Although most seals have a thin layer of fur, it is their blubber that truly keeps them warm. They spend most of their time in the water; they can even sleep underwater and come up to breathe without waking up! However, like walruses, they need to spend time on ice floes or land to rest or give birth.

Elephant seals get their name from the male's trunk-like nose.

The only way for people to get to the true North Pole is on a ship. This ship is a type called an icebreaker.

POLAR LIFE AND EXPLORATION

Chapter Four

No humans live at the exact location of the North or South Pole as of 2025. The communities that live near the North Pole make their home in the Arctic tundra. In the South, scientists live on Antarctica, but not at the exact location of the South Pole. However, people have visited both poles.

Northern Living

At the North Pole, the pack ice moves in the water, shrinks in the summer, and grows in the winter, which makes it impossible to build a permanent house. Furthermore, there is nothing around for miles. A person could hunt for food, but there is no wood or coal to make a fire for warmth or cooking, no stores to get supplies, and no one to trade with.

> **Fast Fact**
>
> In an emergency, a person could eat snow to get the water they need. However, this is not a long-term solution. Eating cold snow increases a person's risk of getting **hypothermia**.

The North Pole is surrounded by water, but none of it is potable, or drinkable. A person would have to remove the salt from the ocean water before they could drink it, which would take a lot of time and energy. For all of these reasons, people have not made communities at the geographic North Pole. However, about 4 million people live in the Arctic tundra near the North Pole.

Life in the Arctic can be hard. There are not many roads, and there are frequent snowstorms in the winter, so travel is often difficult. When people do go out, they need to dress very warmly. Indigenous, or Native, communities in the Arctic have been hunting animals such as seals for their warm fur for thousands of years.

Longyearbyen in Norway, shown here, is the northernmost town on the planet. It is about 621 miles (1,000 km) from the North Pole. The few thousand people who live in the town generally get around on snowmobiles, and they carry guns to keep themselves safe from polar bears, which are more aggressive than any other kind of bear.

THE FIRST EXPLORERS

The first person to travel to the geographic South Pole was a Norwegian man named Roald Amundsen. His team calculated where the pole would be and reached that spot in 1911. Amundsen had originally planned to be the first person to visit the North Pole. However, when he heard that it had been done by someone else in 1909, he switched his plans to reaching the South Pole.

The person Amundsen heard about was an American named Robert Peary. However, the North Pole discovery is a little more complicated than the South Pole discovery. After Peary announced that he had found the North Pole, researchers discovered that his calculations might have been wrong. This means he may have been near, but not exactly at, the geographic North Pole. However, he and his fellow explorer, Matthew Henson, still generally get credit for the discovery.

In Antarctica, there is a **ceremonial** South Pole marker. It is surrounded by the flags of the 12 countries that first created the agreement to share access to Antarctica.

They also hunt and raise animals for food. Many communities now have at least one grocery store, but the cost of getting the items so far north tends to make the items very expensive.

Fast Fact

Antarctica is one of the only places in the world that has never had an Indigenous population. This means it is one of the only places to have truly been discovered by modern explorers, rather than taken from people who were already living there.

Southern Living

No single country owns Antarctica. Instead, multiple countries control it together, and they use it mainly for scientific research. The only settlement there is for the scientific community.

Antarctica has about 66 scientific bases. Nearly 40 of them are lived in all year. The rest shut down in the winter. The largest base on the continent is McMurdo, which is the one that belongs to the United States.

Fast Fact

About 4,000 people live in Antarctica in the summer, and only about 1,000 in the winter. In the summer, in addition to scientists, tourists from cruise ships visit the continent. Between 2018 and 2019, more than 55,000 tourists visited Antarctica—the highest number in history.

McMurdo is open all year round, so the scientists who live there need to be careful on days when the weather is dangerous, or unsafe. For example, sometimes wind whips up the snow, making it easy for someone to get lost just a few steps from their own door. Antarctic residents generally

This picture of McMurdo Station was taken from on top of a nearby hill.

make sure they have food in their room for such days so they don't have to worry about getting lost on their way to the cafeteria.

Life is extreme at both ends of the earth, but that doesn't stop people, animals, and plants from living and even thriving there. With the right adaptations, anything is possible!

THINK ABOUT IT!

1. Why do you think the magnetic North and South Poles drift?

2. How might polar bears be affected by global warming?

3. Why do you think there are penguins at the South Pole, but not the North Pole?

4. What other challenges do you think come with living in the Arctic?

GLOSSARY

adaptation: A change in a living thing that helps it live better in its habitat.

aggressive: Acting with forceful energy and determination.

ceremonial: Only for show; not reflecting reality or serving any functional purpose.

circulatory system: The system that pumps blood throughout the body.

describe: To give an account of something in words.

evolve: To change slowly over time.

hypothermia: Reduction of the body temperature to an abnormally low level.

iconic: Widely recognized and well-established.

intense: To an extreme degree.

oriented: Functionally directed or arranged.

perpendicular: Forming a right angle with a given line or plane.

sparse: Few and scattered, not thickly grown.

thrive: To grow and live well.

tundra: A treeless plain, especially of Arctic regions, having a permanently frozen layer below the surface soil and plant life made up mostly of mosses, lichens, herbs, and very small shrubs.

FIND OUT MORE

Books

Carmichael, L.E. *Polar: Wildlife at the Ends of the Earth*. Toronto, ON: Kids Can Press, 2023.

Kington, Emily. *Polar Regions*. Cornwall, UK: Hungry Tomato, 2021.

Prior, Jennifer Overend. *The Polar Regions*. Huntington Beach, CA: Teacher Created Materials, 2023.

Websites

Kidcyber: North and South Poles
www.kidcyber.com.au/north-and-south-poles
Learn more about the difference between the geographic and magnetic poles.

Sea History for Kids: The North and South Poles
seahistory.org/sea-history-for-kids/the-north-and-south-poles-like-no-place-else-on-earth
Read more about the discovery of each of the poles.

YouTube: Surface Area to Volume Ratio Explained
www.youtube.com/watch?v=huKUJsqik2I
This video explains why larger creatures have a smaller ratio of surface area to volume.

Publisher's note to educators and parents: Our editors have carefully reviewed these websites to ensure that they are suitable for students. Many websites change frequently, however, and we cannot guarantee that a site's future contents will continue to meet our high standards of quality and educational value. Be advised that students should be closely supervised whenever they access the internet.

INDEX